Red Tracers;
The 482nd Anti-Aircraft Artillery Battalion
in WWII

Red Tracers
The 482nd Anti-Aircraft Artillery Battalion in WWII

by

Master Sergeant Walt Cross
U.S. Army, Retired

Author of

Flash – Bang
The Unit History of the 285th Field Artillery (Observation) Battalion

The Fastest Gun in the Big Red One
1st Battalion, 7th U.S. Artillery in Vietnam 1965 - 1970

Custer's Lost Officer;
the Search for Lieutenant Henry Moore Harrington, 7th U.S. Cavalry

Editor of

From the Beaches to the Baltic
The Story of the 7th Armored Division in WWII

Tank Busters
The 607th Tank Destroyer Battalion

Dire Wolf Books are available online at
www.lulu.com/greenpheon7

Dire Wolf Books, an imprint of:

Cross Publications
Stillwater, Oklahoma

ISBN: 978-0-9850996-6-4

502 E. Liberty Avenue
Stillwater, Oklahoma 74075

www.lulu.com/greenpheon7

Printed in the United States of America
First Edition

The unit history of the 482nd Anti-Aircraft Automatic Weapons Battalion was not written or published at the end of WWII. The hope of the author is that this history, combined from unit documents, has captured the spirit of the unit history bookletss written at the end of the war.

The front cover design includes an adapted illustration from page 44 of the 1994 ADA Yearbook and is believed to be in the public domain. Front Cover design by Walt Cross.

Acknowledgments

This book would not have been possible without the material and moral support of Mr. Jim Lefemine, a history writer in his own right, and retired Army Staff Sergeant Charles (Chuck) J. Gregorovich, a member of the greatest generation, and a veteran of Battery A, 482nd AAA AW (SP) Battalion.

I am grateful to both of these men for providing vital information and original documents from which this unit history was written. I also want to acknowledge that without soldiers like Staff Sergeant Gregorovich, we would not have inherited the precious freedoms that so many Americans now seem to take for granted.

Dedication

This book is dedicated to Private First Class Earl L. Cross an armored cavalryman of Company F, 87th Cavalry Reconnaissance Squadron (Mechanized), 7th Armored Division. He has always been my hero.

Walt Cross
Master Sergeant
U.S. Army Retired

Dad in Paris, spring 1945

Table of Contents

The Need for Air Defense	1
France and Luxembourg	10
St. Vith, Belgium	18
Bastogne, Belgium	29
Longvilly and Bastogne	55
Mersch, Luxembourg	73
Germany	87
To Save a Bridge	90
The Final Push	121
Occupation Duty	123
Presidential Unit Citations	130

Enemy Aircraft 136

One Soldier's Story 138

Officer Roster at War's End 143

Author's Afterword 146

Unofficial Unit Insignia 148

About the Author 149

The Need for Air Defense

The German Luftwaffe was one of the most potent of the enemy's weapons from the very opening of the European Theater of war. Over-night the Stuka dive bomber was catapulted into a terror weapon. The Allied air forces were behind the Germans in aircraft design and rapidity of production.

The U.S. War Department, precursor of today's Department of Defense, quickly began to create a large anti-aircraft branch of the artillery. This was a comprehensive plan that included not only units for fixed defense positions such as along the U.S. coastline, but also mobile units.

For the ground forces the basic anti-aircraft unit was to be a battalion sized element. This unit would consist of about forty officers and eight hundred men. Commanded by a lieutenant colonel, these battalions would be armed with various sized calibers of guns and machineguns mounted on a variety of military vehicles.

Tent housing for soldiers at Camp Hulen, Texas. Photograph courtesy Private First Class Carl W. Leach of the 435th AAA AW Battalion.

There are indications that the 482nd began its military life, like many other air defense units, as a coastal artillery battalion. The order below seems to bear this out as the 482nd is reorganized in May of 1943.

At that time Lieutenant Colonel C. J. Girard was in command of the battalion.

HEADQUARTERS
482d AAA Auto Wpns Bn (SP)
Camp Hulen, Texas

May 10, 1943

GENERAL ORDERS)
:
NUMBER--------2)

1. Reorganization of 482nd AAA Auto Wpns Bn (SP).

In compliance with restricted letter file AG 320.2 (4-28-43 OB-I-GNGCTM) War Department The Adjutant General's Office, Washington, D. C. dated May 1, 1943, Subject: Redesignation and Reorganization of Coast Artillery Battalions (AA); and par 1 GO 40 HQ AA Command Richmond, Va. dated May 3, 1943; and par 2, restricted immediate action letter AAC 322.7 K-85-GNSTO (5-1-43 HQ AA Command, Richmond, Va. dated May 4, 1943; and par 1 GO 23 HQ AAATC, Cp Hulen, Texas dated May 7, 1943, this battalion is reorganized under Tentative Table of Organization #44-75, effective this date. In effecting tentative reorganization under the new table, no Non Commissioned Officer will be reduced in grade. Surplus created in any Officer, Warrant Officer or Enlisted Grade as result of this reorganization will be carried until transferred to other Organizations.

2. Reorganization of Staff: The staff of this battalion is reorganized as follows:

Executive Officer		Major Ray E. Leggett	0366909
Plans and Training Officer	S-3	Capt. Francis B. Postelle	0360252
Intelligence Officer	S-2	1st Lt. Jonathan O. Nottingham	0280956
Liason Officer		1st Lt. William F. Varney	01043821
Adjutant	S-1	2nd Lt. James K. Ness	01050833
Supply and Munitions	S-4	2nd Lt. James A. Hansen	01051168

3. General Orders No 1 was published under designation of 482nd CA Bn (AA)(Auto Wpns) Sem which was effective February 20, 1943.

C. J. Girard
C. J. GIRARD
Lt. Col., C.A.C.
Commanding

DISTRIBUTION
1 Each Officer Concerned
1 CO each Battery
1 201 File each Officer Concerned
3 CO 23rd CA GP
5 CO Cp Hulen, Texas
5 CG AAATC, Camp Hulen, Texas
5 CG AA Command, Richmond, Va.
5 T.A.G.
10 File

Four months after this order, Lieutenant Colonel Vincent P. Lupinacci assumed command of the battalion. It was Lupinacci who directed the training of the 482nd AAA, deployed with the unit, and commanded them while overseas.

Lupinacci, the son of Sicilian immigrants, was a reserve officer from Pennsylvania and worked for the Bell Telephone company before and after the war. During the 482nd AAA AW (SP) Battalion's time in the European Theater, Lupinacci would fight to keep the anti-aircraft role of his unit in the forefront of his operations.

There is a saying in the Army, that the best plan of action rarely survives first contact with the enemy. Here is the actual organization of the 482nd Anti-Aircraft Artillery (Automatic Weapons) Battalion.

Officers	36
Warrant Officers	02
Enlisted Men	688
Total Strength	726

Headquarters Battery

Officers	09
Warrant Officers	02
Enlisted Men	122
Total Strength	133

Vehicles for Headquarters Battery consisted of five unarmored half-tracks, twelve ¼ ton trucks (jeeps), thirteen 2½ ton trucks (deuce and a half), and one heavy wrecker truck.

There were four letter gun batteries designated A, B, C, and D. Each battery had six officers and 141 enlisted men for a total of twenty four officers and five hundred sixty four enlisted men.

Vehicles for each battery included one armored halftrack, eight M-15 self-propelled gun mounts, eight M-16 self-propelled gun mounts, four jeeps and two deuce and a half's. This provided the battalion with sixty four armored self-propelled gun mounts. Below is an M-15 half track with two air cooled .50 caliber machineguns and a 37mm main gun.

The 482nd AAA AW (SP) Battalion departed their training base at Camp

Hulen, Texas[1] on May 25, 1944. They arrived by truck at Camp Barkley near Abilene that same day. Processing for overseas assignment continued until the battalion departed again, this time by troop train on 1 August 1944 bound for Camp Kilmer, New Jersey.

They arrived at Camp Kilmer on the 4th of August and six days later boarded the *U.S.S. Hermitage* on 10 August and left from the New York Harbor the very next morning.

They arrived in Liverpool, England on the 23rd of that month and were sent to camp at Blackshaw Moor near Leek. After three weeks in the United Kingdom the battalion left at Southhampton harbor about the *Daniel Heister* on September the 18th.

[1] Camp Hulen was located midway between Houston and Corpus Christi, Texas.

An M15 lands on a French beach.

An M-16 halftrack with its four guns.

France and Luxembourg

The 482nd AAA AW (SP) Battalion landed on Utah Beach in France on the 23rd of September, 1944 and after bivouacking near Blosville, moved out to meet the 9th Armored Division in Luxembourg as an attached (not joined) unit.

Upon arrival on November 2, after an uneventful journey, the battalion took up defensive positions around the town of Mersch.

The placement of the batteries was determined by the likely avenues of an enemy attack, and the relative importance of the defense of a particular position or unit.

The weather was rainy with occasional snow squalls that did not amount to any significant accumulation.

The area was, of course, muddy all the time.

The battalion was under the command of Lieutenant Colonel Lupinacci with six captains in command of the subordinate batteries and detachments as follows.

Captain Benjamin J. Herman, Battery A.
Captain Carlton G. Denton, Battery B.
Captain Denny J. Lovoi, Battery C.
Captain Harry A. Ripkey, Battery D.
Captain Edward J. Noone, HQ's Battery.
Captain Jack Viscofsky, Medical Detachment.

Battery A was to defend the 3rd and 73rd artillery battalions located near the villages of Consdorf and Haller respectively.

Battery B was attached to Combat Command B of the 9th Armored near the town of Trois Vierges and deployed as directed by that commander.

Battery C was in defense of the 9th Division's command post, the railhead and the bridge at Mersch. To augment Battery C, three 40mm Bofers guns and three M-51 mounts, all from the 635th AAA (AW) Battalion were attached.[2] But these additional guns were soon withdrawn and used elsewhere.

[2] A 40mm Swedish made anti-aircraft automatic, multi-purpose gun. An M-51 was a trailer mount for anti-aircraft guns.

A trailer mounted M-51.

A Bofers gun.

Battery D was sent to defend the division *trains* comprised of administrative and supply vehicles near the town of Hollenfels.

Changes of these dispositions were made by the 9th Armored Division's G-3 (operations section) as needed. Noteworthy among these changes was the assignment of D Battery to the defense of the 811th Tank Destroyer Battalion near Scheidgein. The 811th TD Battalion was the subject of my book *Tank Busters 2* which readers of this narrative may find interesting.

During this time frame no position or unit under the protection of the 482nd AAA AW Battalion suffered any loss or damage from enemy aircraft.

At 1530 hours on November 6 two ME-109s strafed the positions of the 3rd and 73rd artillery battalions. One of the planes was shot down by Battery A, resulting in the first awarded (confirmed) 'kill' for the battalion. Neither battalion suffered any damages or casualties.

During the period of 25 to 30 November many V1 flying bombs passed over the 9th Armored Division but not close enough for the battalion to engage.

To counter the wet weather certain changes were required to the battalion's SOP (standard operating procedures) regarding field fortifications. The normal procedure was to 'dig in' tracked vehicles, but this proved untenable as the positions were soon filled with water.

An AA halftrack 'dug-in'.

Instead of *digging in*, revetments were constructed around halftracks and other positions comprised of cut wooden poles reinforced with turf and camouflaged to match the terrain.

It was found that tracked vehicles should not be placed too near rivers or streams. Too often rising water levels required the movement of the vehicles further from the flowing water. So the time the battalion spent near Mersch was not wasted as lessons were learned.

The battalion also experienced some communications problems primarily because radio silence was strictly adhered to and the movement of personnel after dark was prohibited by the division.

Eventually the battalion procured some field telephones and enough land wire to connect the gun sections to headquarters to receive fire missions.[3]

Most of the aerial activity was V1 bombs, so the battalion was assigned the task of keeping a count of these robot aircraft.

During November the battalion expended two thousand and forty five

[3] The author served in Battery A, 1st Battalion 7th Artillery during the Vietnam War. The section that provided firing solutions to the 105mm guns of the battery was called the FDC or fire direction center.

rounds of .50 caliber ammunition and twenty one 37mm high explosive shells.

Battle of the Bulge, St. Vith

The orders that Combat Command B, 9th Armored Division received on the 13th of December 1944 demonstrates just how far out of touch Eisenhower's operations and intelligence sections were regarding the enemy's intent.

On that day, just three days before the explosion of the gigantic struggle that would come to be known as the 'Battle of the Bulge', Combat Command B was ordered to support an attack by the 28th Infantry Division.

In the coming days the 482nd AAA AW Battalion and indeed the 9th Armored Division, would be split into two or more elements and fight in the major roadblock Belgian towns of both St. Vith and Bastogne. The battle fought in these two city-towns would determine the fate of Hitler's Third Reich.

Battery B of the 482nd was tasked to defend the assembly area near Faymonville, Belgium for the 9th Armored Division units slated to support the attack.

But events were to dictate otherwise, and at 0100 hours on the 17th of December Battery B was ordered to provide Anti-aircraft protection to the 16th Field Artillery as they moved to Breidfeld, Belgium. They did not encounter enemy aircraft but did receive some 88mm shell fire from the Germans.

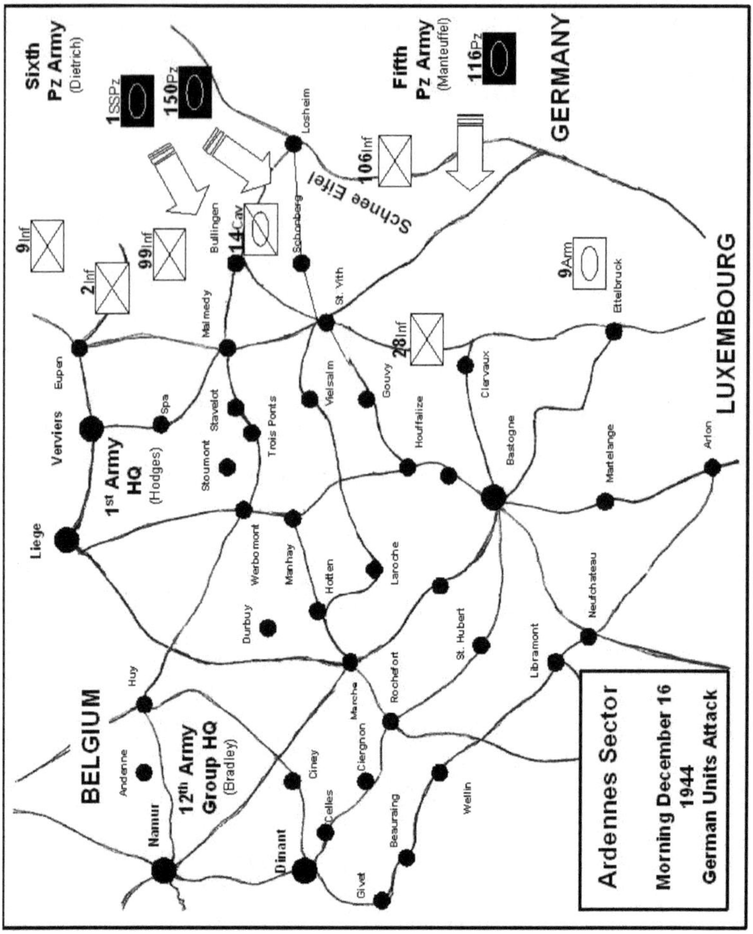

Ardennes Sector December 16th.

At 0550 hours that same day the battery received a message from Combat Command B that enemy armor was

pushing its way toward St. Vith, a vital road and railroad junction town some distance to the east. At that time it was also the headquarters for the doomed 106th Infantry Division.[4]

Captain Denton, the battery commander, assigned his First Platoon under Lieutenant Winchester to accompany the 14th Tank Battalion in a thrust toward St. Vith.

He then assigned his 2nd Platoon with Lieutenant Haick in charge to support Company C, 811th Tank Destroyer Battalion. Together with a platoon from the 27th Armored Infantry Battalion, this 'scratch' taskforce set out for St. Vith at 1800 hours.[5]

---

[4] Two of the three regiments of the 106th Infantry Division surrendered after being surrounded on the Schnee Eifel ridge east of St. Vith, Belgium.

[5] The author's father, PFC Earl L. Cross of Company F, 87th Cavalry Recon Squadron, 7th Armored Division, would arrive in St. Vith on the afternoon of this same day. His M5A1 Stuart tank was the first 7th AD vehicle to arrive. The 87th Cavalry would receive the Presidential Unit Citation.

Arriving at the threatened town this taskforce moved to a position one mile north of the already embattled city. At this point the 2nd Platoon and C Company of the 811th Tank Destroyer Battalion rolled north to a position between Malmedy and St. Vith to repulse an enemy attack of infantry and Mark III panzers.[6] Contact was made with the Germans and three panzers were destroyed by the combined arms of the 811th tank destroyers and the 482nd Anti-Aircraft Artillery.

---

[6] Malmedy was the location of the infamous "Malmedy Massacre" of December 17, 1944. The author has written a unit history of the 285th Field Artillery (Observation) Battalion that was virtually destroyed in Malmedy titled *Flash – Bang the Unit History of the 285th Field Artillery (Observation) Battalion.* The German unit involved was an *SS* unit designated Kampfgruppe Peiper. These men were later tried for their war crimes, convicted and imprisoned.

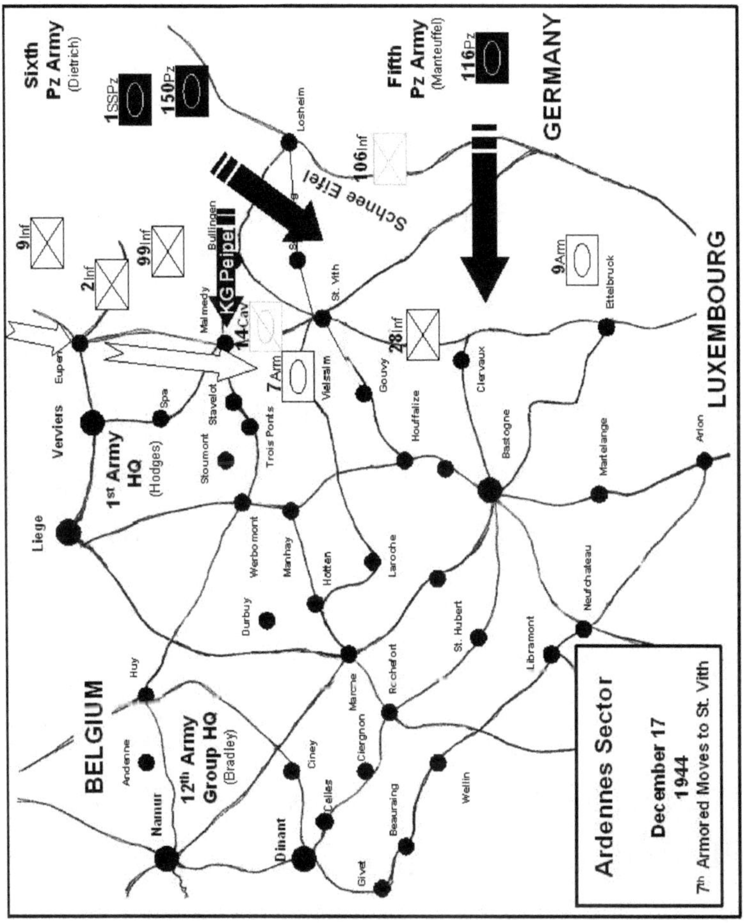

Ardennes Sector December 17.

This seemed to stop the enemy advance toward St. Vith, but in truth this German SS taskforce designated Kampf-gruppe Peiper, was headed west towards

Antwerp and its seaports where Allied supplies were being landed. Antwerp was Hitler's designated goal and the target for the massive surprise German attack.

On the following morning units of the 7th Armored Division arrived and the 2nd Platoon moved to Breidfeld, Belgium where the rest of Combat Command B, 9th Armored Division was under enemy artillery fire.

Private Cerda of the 1st Platoon was wounded during this shelling which fell all day on December 19th. At 1850 hours Combat Command B moved from Breidfeld to the village of Grindhausen where Battery B was assigned to support the 27th Armored Infantry Battalion.

On the 20th of December heavy German shelling continued against all units of Combat Command B in preparation for a general attack. At 1400 hours enemy panzers supported by infantry came against the right flank of the command from the southwest. The entire command engaged the advancing Germans who could be seen as they came over the surrounding hilltops. Battery B opened up with every halftrack's heavy machineguns and the resulting outpouring of lead and red tracers completed the repulse of the Germans.

When the Germans pulled back the enemy shelling returned and continued throughout the night and into the following day.

On the afternoon of the 21st German infantry in mass moved against the left flank coming directly from the east. Although turned back by Combat Command B to the north the enemy broke through elements of the 7th Armored Division.

B Battery was soon ordered to advance to the north in support of the 27th Armored Infantry Battalion. At 0900 hours Captain Denton's jeep came under heavy small arms fire. He pulled his jeep off the road at the command post of the 27th Armored Infantry Battalion and found that it too was under attack.

Jumping from his vehicle Captain Denton mounted a nearby halftrack and engaged the enemy near the command post with the halftrack's .50 caliber machinegun. He then contacted the battery by radio and brought them into the attack. The halftracks quickly arrived and inflicted heavy losses on the attacking Germans.

Entering the 27th Armored Infantry Battalion's command post Captain

Denton was temporarily taken prisoner by German soldiers who had captured the command post. However, as the situation stabilized and the Germans were cut off, he was soon released. All other remaining enemy soldiers were either captured or killed.

During the action Lieutenant Jackson was wounded in the leg by gunfire and Tech 5 Macario Torres braved heavy enemy fire and came to his aid.

At 2200 hours the enemy closed in on Combat Command B from all sides. Under continuous enemy shelling the entire command pulled out and relocated overnight and into the morning of the 23rd to the vicinity of the town of Vaux Chevanna, Belgium. The battle continued all the way with hot firefights erupting around the command's perimeter. The enemy found every step they took contested by the men of the 9th Armored Division.

By 0945 hours on Christmas Eve, the command moved on to Bras, Belgium and the battery set up for air defense. At 1400 hours an ME-109 flew into the area diving and strafing at low altitude. The battery engaged the hostile aircraft but lifted its fire when a P-51 Mustang was seen in pursuit.

Under cover of darkness Combat Command B left Bras at 2030 hours and moved to Xhoris, Belgium. Battery B continued to provide air defense cover, but was soon placed in a reserve status.

During the Defense of St. Vith, Battery B inflicted more than five hundred enemy casualties with its accurate fire guided by red tracers from machineguns and 37mm cannon. During the action it was noted that when the multiple guns of the battery erupted and the streams of red tracers began, enemy fire ceased.

During the fighting the battery expended 150,000 rounds of .50 caliber ammunition and 1200 rounds of 37mm shells. Amazingly there were no killed in action and no loss of equipment.[7]

Bastogne, Belgium

On the morning of December 16, 1944 Battery A of the 482nd AAA AW Battalion was in defense of the 811th Tank Destroyer Battalion near the town of Scheidgein here is what happened according to the unit history of the 811th.

[7] For a more in-depth look at the important battle for St. Vith see the author's book *From the Beaches to the Baltic, the story of the 7th Armored Division in WWII.*

...at 0600 hours on 16 December 1944 the Germans launched a massive artillery barrage with a large amount of shells falling in the Consdorf and Scheidgein area. The observation posts at Berdorf and Lauterborn, in the processes of moving, returned to their positions.

However the Lauterborn OP was untenable due to intense enemy fire. Captain Samuel Parson returned to Berdorf at 0800 hours to discover the enemy was attacking the town with considerable strength.

Both the 811th and the 3rd Field Artillery Battalion were acting as division artillery. Company A of the 89th Cavalry Reconnaissance Squadron was patrolling in the area of Beaufort providing a screen along the Sauer

River. Meanwhile the 60th Armored Infantry Battalion was generally located along the river between Reisdorf and Beaufort holding a defensive position. This force was under Combat Command A, 9th Armored Division.

The 811th unit history continues the story.

...Company B left at 1000 hours and continued toward its new position at Haller. However, the S-2 (intelligence officer) reported that the enemy had attacked all along the VIII Corps frontline and that the towns of Berdorf, Lauterborn, Osweiler and Dickweiler were held by the enemy. Reconnaissance in the vicinity of Beaufort to the north of Company B's new positions was halted due to heavy artillery fire and enemy infantry action.

There was heavy fighting all along the Sauer River with German armor and infantry trying to force their way across the river even under heavy American fire.

The following day the Second Platoon of Battery A and Company B of the 811th Tank Destroyer Battalion came under heavy enemy artillery fire. One M-16 halftrack of the Second Platoon was hit and suffered damage and three of its crew were wounded. The intense barrage fire forced the vehicles of both units to evacuate to the Christnach - Savelborn road.

In nearby Haller the First Platoon also came under artillery fire but remained in position.

Under pressure of the enemy the 482nd Battalion command post was moved from Christnach back to the town of Fels and a forward battery CP at Haller, to which Battery A was moved in support of the First Platoon. So the battery moved out while continuing to take enemy fire. The entire battery now prepared for ground defense against an advancing and determined enemy.

Maintaining this concentration of artillery fire, the German infantry took position of a densely wooded ravine directly east of the town of Christnach. The ravine continued south of the town and provided the Germans cover all the way back to the village of Echternach on the Sauer River. This posed an immediate enemy threat to the flank of the 9th Armored Division as the ravine cut exactly through the boundary line between the 9th Armored and the 4th Infantry divisions.

Meantime the Germans had isolated and surrounded Company A of the 89th Cavalry Recon Squadron as well as the 60th Armored Infantry Battalion in and around the town of Beaufort.

In the ensuing fight these units were effectively neutralized in the defense of Combat Command A's sector.

Although elements of these units would infiltrate back to the Amer-ican lines, many GIs were captured or killed and the units were in grave need of reorganization.

This left only Company A of the 482nd AAA AW in defense of the combat command's sector. So during the night the 3rd Field Artillery Battalion was moved to the vicinity of Fels and B Company of the 811th Tank Destroyer Battalion was shifted to the area of Christnach.

Before dawn on December 18th at about 0600 hours, Company C of the 89th Cavalry Recon Squadron[8] was sent to defend Haller and Battery A was attached to provide artillery support.

Three hours after arriving in Haller these two units were designated a task force and sent to take the high ground north of the town.

Under the command of lieutenants Higgins and Koegan, the men of the 89th deployed as infantry and were supported by the M-16 halftracks while the M-15s and a few M-16s remained in Haller to furnish fire cover with their 37mm guns and machineguns.

Using a narrow ravine the infantry moved under cover to the vicinity of Beaufort, the objective of the planned

[8] Cavalry reconnaissance squadrons were for the most part in light armored cars armed with a 37mm main gun and a .30 caliber machinegun.

assault. Reaching the high ground they had a commanding view of the town.

The infantry came under small arms fire and enemy grenadiers armed with panzerfausts (German bazookas) waited in front of Beaufort.

A German NCO demonstrates a panzer-faust (German bazooka) to students.

The air was filled with zipping bullets, the stutter of U.S. machineguns and the answering ripping sound of the

German MG42 machinegun. The loud underlying 'whump' of 37mm guns punctuated the seriousness of the fight.

Dense woods and limited road access hampered the effective use of the M-16s. The infantry returned the enemy's fire, causing them to withdraw into the town proper. However, after assessing the situation it became apparent the task-force did not have sufficient numbers or the necessary firepower to take Beaufort. The taskforce withdrew back to Haller after killing twenty and wounding fifteen of the enemy.

During the fight a U.S. scout car, captured and put into use by the Germans, was knocked out as well as two heavy mortar positions destroyed by 37mm fire. The taskforce withdrew in good order and the enemy did not initially pursue.

The men of the 89th Cavalry took positions inside Haller as the 482nd AAA AW spread their vehicles on either flank of the armored cars.

For the next two and half hours mortar and artillery shells fell on Haller and its stalwart defenders. As the artillery fire died away German infantry came to the fore to press the attack. But the fire of the quad mounted machine-guns of the M-16s added to the twin .50 calibers and 37mm guns of both the M-15 halftracks and the armored cars of the 89th Cavalry drove them back. By 1800 hours the attack was over. The Germans left behind them twenty dead and twenty five wounded.

During the fight one M-15 received a direct hit by German artillery and was destroyed. The crew wisely took to nearby foxholes during the barrage and suffered no casualties.

The Germans had taken a hard whipping and withdrew, leaving the artillery to continue harassing the Americans in Haller.

A mobile reserve of M-16s, able to move about to the most threatened areas in the middle of the fight; were instrumental in stopping the enemy assault.

That night German infantry forced their way into Haller on the right flank of the defenders and was also attacking Christnach in strength. This attack occurred directly behind the Haller taskforce, placing them in grave danger of being overrun. As if that were not enough, enemy forces penetrated the area west of Haller and drove as far as the village of Flechette. The defense of

Haller became near impossible as attacks were launched to the front and both flanks of the taskforce.

Artillery fire was received from enemy guns in Beaufort and mortar shells rained down from both flanks. However, the intrepid men of the taskforce held their ground until 0300 hours the morning of the 19th of December. Then they were ordered by Combat Command A of the 9th Armored Division to fall back and form a defensive line on the road stretching between Christnach and Savelborn.

During the withdrawal one track became bogged down so badly it could not be moved and was destroyed in place. The crew took position and engaged the Germans with small arms fire while others slit the tires, destroyed the machineguns and detonated a thermite grenade on the engine. It was all completed without the loss of a single man.

In the new defensive position A Battery was still susceptible to artillery fire, but the threat of being overrun was removed.

Company B of the 811th Tank Destroyer Battalion, augmented with some elements of their own Recon Company, denied the town of Christnach to the enemy pushing in from the east.

Battery A, 482nd AAA Battalion held the defensive line on the road as mentioned while the town of Savelborn was defended by Company C of the 89th Cavalry Recon Squadron.

In the meantime, the 60th Armored Infantry Battalion was reorganized and held a defensive line from Savelborn to the town of Schieren supported by Company A of the 89th Cavalry.

The road that Battery A of the 482nd occupied; lay parallel to a ravine and along the slope of a long, treeless, and marshy hill. In the ravine at the foot of the hill were woods, a single house, and the enemy. The halftracks were spread along the road for two thousand yards with an excellent field of fire into the ravine, the woods, and the slope of the hill. The battery's command post was in a farm house midway between the two towns of Savelborn and Christnach.

Captain B.J. Herman, Battery A's commander, was a trained infantry officer. He recognized that the enemy occupied the ravine and so he organized a recon patrol under Lieutenant James French along with Corporal Piazza and

Private West to reconnoiter and make visual contact with the Germans.

At 1000 hours on the 19th, the patrol set off with Lieutenant Keogan in the loft of the command post, observing their progress and if needed, providing supporting fires from the halftracks.

The house proved to be an enemy strongpoint and when the Germans detected the patrol they soon pinned them down by fire. Lieutenant Keogan ordered the nearest squad to return fire and the German machinegun fell silent. The patrol beat a hasty retreat and reported what they had seen.

At 1150 hours fire was brought upon the enemy positioned in the house from the 3rd Field Artillery. After the fire lifted Captain Herman sent a combat patrol consisting of Sergeant Carreson with corporals Canipe, Brown, and

Huggins and an additional three Tec 5's named Stoles, Wheeler and Osborne.

Under supporting fires these men attacked the enemy occupied house with hand grenades and small arms fire.

In the exchange Tec 5 Wheeler was wounded in the stomach by enemy return fire but continued on and attacked the nearby barn.

Kicking the barn door open Wheeler entered it firing and killing three enemy soldiers using his .45 caliber Thompson sub-machinegun. The ferocious attack on the house and barn was too much for the Germans, and they surrendered. Amazingly the seven man patrol captured fifty-nine enemy soldiers and their officer!

The 811th unit history states that:

On the morning of the 18th of December Sergeant Shnell was sent to make contact and act as liaison with elements of the 4th Infantry Division on the battalion's right (south) flank. The 482nd Anti-Aircraft Artillery Battalion protected our left flank and on the 19th Lieutenant Horning of Company B joined with that unit for an attack on a farmhouse occupied by the enemy. The joint attack killed 40 Germans and resulted in the capture of 61 grenadiers of the 276th Volks Grenadier Division...

Prisoners...reported that our attack on Mullerthal had killed half of their infantry company...and both their regimental and division commanders...

Later interrogation of the captured Germans revealed their great fear of the M-16 and M-15 fire and that they believed they were some kind of ‘secret weapon’.

For the rest of the day the battery fired intermittently into the woods and house to keep the enemy at bay.

The night of the 19th and morning of the 20th of December the battery continued to send out patrols that made contact with the enemy and kept up small firefights.

During the day of the 20th the battery continued to skirmish with the enemy and received heavy concentrations of artillery and mortar fire, but doggedly maintained their position.

Captain Herman kept it warm for the Germans too by directing the 3rd Field Artillery’s fire missions on the woods.

However, the Germans continued their relentless artillery barrages directed at the battery’s position.

As the sun rose on the morning of the 21st of December, a taskforce was assembled and moved out at noon. The force consisted of the usual units and elements of the 60th Armored Infantry Battalion, the 811th Tank Destroyer Battalion, and a platoon of tanks from the 19th Tank Battalion.

Battery A of the 482nd supported this force with six halftracks. Three M-16s were under the direction of Lieutenant Koegan and the other three under that of Lieutenant French.[9]

The M-16s advanced and encountering German reinforce-ments on the Berdorf – Waldbillig road inflicted a number of casualties on them and stopped the German movement into Waldbillig. They then followed the advancing tanks into the town and

[9] The 811th Tank Destroyer history says this force was designated Taskforce Brownfield.

engaged enemy strongpoints. Enemy casualties from their fire included thirty-five dead and twenty wounded.

Battery A suffered two men wounded by small arms.

...after intense fighting, some of it house to hour, Waldbillig was taken. The cost was high for the 10th Armored who lost two tanks and a company of infantry that was killed, wounded, or missing in action.[10]

Unable to hold Waldbillig because of increasing enemy pressure, Taskforce Brownfield along with other nearby units, withdrew to positions to the southwest. There, Battery A occupied some well defiladed positions on the high ground.

---

[10] 811th Tank Destroyer Battalion unit history.

The situation now became static with continuous enemy mortar, artillery and *nebelwerfer* fire.[11] The Germans continued to move on Waldbillig and strengthened their furious attacks.

The next morning at 0700 hours 22 December, the six tracks were returned to the control of the battery commander. The battery then set out once again to defend the Savelborn – Christnach road.

Because the area on either side of the road was a marsh, the battery had to remain on the road itself in view of the enemy gunners. However, the battery determinedly stayed at their stations and returned fire whenever the enemy dared to show themselves.

Some halftracks found protection near or behind barns and hay stacks while others simply occupied the open

[11] Rocket fire, sometimes referred to as 'screaming meemies'.

road, their guns trained in the enemy's direction.

At 1500 hours the Germans mounted an infantry attack that came under determined and heavy fire from the battery. The attack failed, with the enemy sustaining twenty killed and ten wounded for their pains. In return the 482nd AAA AW suffered one casualty from rifle fire.

That night the battery enjoyed supporting fires from the 423rd Field Artillery Battalion augmenting the fires from the 3rd Field Artillery. They were also joined by a platoon of tanks stationed in Savelborn.

The tanks were careful not to move around too much as movement always resulted in heavy shelling by the Germans.

In fact the shelling became so hot that a forward observer team made up of Sergeant Savage and privates first class Brengard and Marks were sent out to spot the enemy artillery's position.

With the information gathered by these men, both the 423rd and 3rd field artillery battalions fired much more accurate and concentrated fires on the German positions. One tangible result that quickly became evident was the destruction of an enemy ammunition dump that went up with a tremendous explosion and continued to 'cook off' shells and ammunition most of the day.

That night for five hours beginning at 0100 hours and lasting until 0600 hours the morning of the 23rd of December the Germans attacked relentlessly. Each time the enemy was repulsed with staggering losses. A single *panzerfaust* team penetrated the battery's line, but two of them were killed and another wounded and captured. An enemy tank attack upon Savelborn was also repulsed and one panzer was destroyed and left burning.

The next afternoon at about 1400 hours a mortar barrage destroyed one of the battery's jeeps and one track killing three men. One of those killed was Corporal Canipe who had distinguished himself during the December 19th attack. Later, at 1800 hours more Battery A men were wounded by artillery fire and medically evacuated. It had been a tough day for the battery.

At 1100 hours on Christmas Eve morning, the battery moved forward in support of an infantry attack to the north of the road. All halftracks moved out, both the M-16s with their quad machineguns and the M-15s with their twin machineguns and the heavier 37mm cannon. Fifty enemy soldiers broke out of the woods trying to escape, but all were killed by fire from Corporal Majerczak who was manning an M-16 gunmount.

On Christmas day the battery was sent to the rear for a refit with ten of their combat vehicles no longer operational. The break was almost like getting a Christmas present.

During this period of action from December 17 – 25 Battery A of the 482nd Anti-Aircraft Artillery Battalion suffered three men KIA[12] fifteen WIA and lost ten halftracks damaged or destroyed in combat. The battery fired 225 rounds of 37mm armor piercing rounds and an additional 325 high explosive shells;

[12] Killed In Action and Wounded in Action.

59,984 rounds of .50 caliber machinegun ammunition, 2,467 rounds of .30 caliber M1 rifle ammo, 1,987 .45 caliber rounds, 870 M1 carbine rounds, and 52 hand grenades expended in combat.

This action by the men manning the battery's halftracks was supported by the battery rear echelon led by First Sergeant Crabbe from the town of Fels to the scene of the fighting. Sergeant Burns, the mess sergeant, was wounded while attempting to provide the men at least one hot meal per day.[13]

The officers of Battery A, Captain Herman; and lieutenants Smith, Higgins, Koegan, Barns and French all demonstrated excellent work in the various tactical roles they played not only as

[13] In 1968 to 1970 when the author served in Vietnam with Battery A, 1/7th Artillery, 1st Infantry Division, this same ratio of one hot meal per day was maintained by the mess section and flown in by chopper. So the efforts of Sergeant Burns and First Sergeant Crabbe were exceptionally good.

anti-aircraft officers but also their leadership as infantry and artillery officers as well.

Longvilly and Bastogne

On the morning of the 17th of December two gun sections of the First Platoon, Battery C, 482nd AAA AW (SP) Battalion in the vicinity of the town of Weiswampach, Luxembourg came under heavy artillery fire. Later that same day Battery C was ordered to proceed to Oberwampach while providing AA defense for Combat Command R, 9th Armored Division. Darkness fell quickly and the battery operated under blackout conditions with only an incomplete knowledge of the enemy's whereabouts.

What was known was that the Germans were advancing and that was to become clearer on the morning of December 18. Orders soon arrived directing Battery C and other elements of Combat Command R in their vicinity to Longvilly, Belgium a trip of some five miles.

Upon arrival the battery was split up into sections and assigned to separate units. The First Platoon was to defend the 73rd Artillery Battalion and placed one section with each firing battery while they setup their command post in the town of Margaret alongside the 73rd Artillery's headquarters.

The Second Platoon took defensive positions around the town of Longvilly keeping their command post nearby on the road leading to Bastogne. The battery command post was set up in the town itself.

Later in the day, in anticipation of intense enemy action, the battery commander ordered his non-combat vehicles back to the town of Bastogne. Enemy artillery continued its tenacious bombardment of the battery all day and all the following night.

Seeking to cut the road between Longvilly and Margaret, the Germans attacked Margaret with tanks supported by infantry, capturing the town after fierce fighting. They succeeded in overrunning the headquarters of the 73rd Artillery Battalion and the command post of the First Platoon, and captured a gun section. Most of the men from the headquarters and command post were either taken prisoner or killed in action. Two officers and an NCO of the First Platoon, lieutenants Cullifer and Gerber and Sergeant Rudell R. Bryce were reported missing in action.

Eleven men of the First Platoon managed to evade capture and escaped on foot. Taking position in a stand of spruce trees, they found themselves in the middle of a fierce tank and artillery battle that ended when 9th Armored Division soldiers drove the enemy back.

The men then made contact with and joined their comrades. They learned the battle had successfully opened the road for Combat Command R to retreat to Bastogne.

At nearly the same time Combat Command R in Longvilly was also under enemy attack in some force.

Taskforce Cherry of the 10th Armored Division sped as quickly as possible to make contact with what remained of the rest of Combat Command R, 9th Armored Division.

The 9th Armored Division units under Colonel Gilbreath were insufficient to stop an entire panzer division.[14] So he

---

[14] At St. Vith the 7th Armored Division augmented by a combat command of the 9th AD as well as various other scraped together units were facing the brunt of an entire panzer army.

split his available forces into three groups and still maintained his command post at Longvilly.

Battery C was to hold their positions until they could shore up their line with elements of the 10 Armored Division.

When Lieutenant Colonel Cherry of the 10th Armored Division entered Colonel Gilbreath's command post the colonel purportedly said *"You couldn't look any better to me if you were Jesus Christ!"*

All elements of Combat Command R were now ordered to withdraw to Bastogne, but the front of the column soon came in contact with a heavily defended German roadblock.

Moving under the fire of German Tiger II tanks from their left flank occupying the high ground, as well as an intense artillery barrage, Combat Command R immediately attacked.

A German tiger tank.

Manning their halftrack mounted guns; Battery C brought their considerable fire to bear on the enemy position while a taskforce was formed to assault the roadblock.

The enemy was well concealed making it difficult to find where the firing was coming from. In response Battery C continuously hammered all likely locations for enemy gunners with their heavy .50 caliber multiple machineguns. There was no doubt the battery was inflicting casualties on the Germans as their fire slackened and the column was able to move forward on the road.

Sergeant Sklenar wrote later about his and his cooks' experience on the Longvilly road.

On or about the 19th the captain sent word that we were going to make a stand and fight to the last man, and it included the kitchen crew. He also sent word to cook a meal for noon. We were parked on this country road, it was a two-lane, narrow highway in a sort of valley—it became the infamous Longvilly Road for C Battery.

We cooked the meal without incident; we hadn't taken any shell fire for some time but we could hear small arms fire, and it was very close. We all knew we were going to get our first taste of close combat. On the right side of the road was a slight hill about a block to the top and that is the way the Krauts were coming.

On the kitchen truck we had a 50-caliber machine gun on a ring mount and that's a lot of fire power, so again I was all Gung Ho. The box on the gun held about 50 rounds but that wasn't good enough for Big Shot Sklenar. I took a full ammo box and set it on the seat which folded down with the back made of steel; I fed the belt with 250 rounds directly from the box into the gun. The other five guys were laying on the ground ready to start firing their carbines; the only thing I told them was to wait till I started firing before they opened up. To show you how green we were to combat, all five guys were shoulder-to-shoulder right in front of the kitchen truck, one grenade would have killed us all—they should have been spread apart. I've said before, you learn real fast in combat, you never make the same mistake twice.

I remember it was about this time my stomach felt like it was tied in knots I was getting nervous... American GIs started backing up, but firing in hand-to-hand combat, the GI's were going down like mowed alfalfa, they were out-numbered and couldn't stop the Krauts, they were being overrun. I told the guys to make sure they had a clear shot at a German before they fired, but to wait for my order—I had my finger on the butterfly trigger ready to fire. Firing that machine gun with that ring mount, I was pretty high up in the air so I could see down this road for [a] *long way.*

About this time I saw a jeep coming toward us, it was flying. It was Captain Lovoi's driver, he was all excited, as we all were, and he said the captain said to get the hell out if we could, to try to make it to Bastogne if possible. This is where I have to say God stepped in; I know the captain gave the order to get out, but I think the captain had to have some guidance from above. I know that if we had stayed there, we wouldn't have lasted five minutes. This is where C Battery took its first casualties as well as some were captured and some were wounded.

We dumped the chow out on the ground and got both trucks on the road and started rolling like hell. We took a few bullet holes in the tarp on the trucks but nothing was damaged. When the captain's driver said the word Bastogne, it was the first time I heard the name. Little did any of us know we would make history and fight in the greatest battle in which the U.S. Army ever participated.

...in our basic training when we were going to cooks and mess sergeants school, we were told we were soldiers first, then cooks, and that was proven at Longvilly. C Battery, and CCR, along with many others, were trapped on Longvilly road, it was the first time my cooks had fired their carbines at a German soldier and the first time I ever

fired my fifty caliber ring mount in combat...We shot our way out and made it to Bastogne, with our kitchen truck...[15]

Vehicles too damaged to move were shoved to the side of the road; their crews thus dismounted continued on foot and brought their small arms fire to bear on the enemy roadblock. In this manner the combat vehicles of Battery C entered Bastogne, and Captain Denny Lovoi gathered the men on foot and brought them to join the rest of the battery.

Although hard to determine the amount of damage inflicted by C Battery's guns, it is known that the enemy's rate of fire died down and eventually the roadblock was abandoned by the enemy.

---

[15] Most if not all quotes from Sergeant Sklenar appeared in various issues of the magazine ADA (air defense artillery) as letters to the editor.

Soldiers of the 482nd Anti-Aircraft Artillery Battalion aimed their formidable quadruple guns directly at the Germans and fired until the barrels glowed red. Mess Sergeant Charles Sklenar and the other cooks of the anti-aircraft battalion joined in, shooting at the enemy for the first time in their Army careers.[16]

During the day on December 20 Battery C men straggled into Bastogne to rejoin the battery. Later that same day Captain Lovoi led a five man patrol down the railroad track 500 yards to the east to recon any enemy movement.

The patrol consisted of Staff Sergeant Charles Sklenar; Tech 4 Charles Stewart, Tech 5 Basil Mecili, Tech 5 Oyler, and Private Joseph Mann. A heavy fog had

[16] From *Those who Hold Bastogne* pp. 43, 45.

moved in but never the less the patrol was subjected to artillery fire. During one barrage Captain Lovoi was wounded bad enough that he was evacuated that night. Sergeant Sklenar wrote of his experience.

...before the 101st Paratroopers arrived... my captain wanted eight men for a patrol, one of my cooks and myself stepped forward, there wasn't anyone else. No one knew how close the Krauts were, none of us were trained for this but our captain would lead us, and that was good enough for me. We went out a couple of miles and ran into the Krauts, after a brief fire fight, we withdrew without a scratch, but coming back my captain was hit quite badly by shrapnel. He didn't make it back till the war's end...

Another patrol took their place when they returned, led by Lieutenant David Doby, who was also wounded by shrapnel but was not evacuated at his own request. He remained with the battery throughout the siege of Bastogne.

This was to prove a stroke of good fortune because later in the afternoon lieutenants Albert Baley and Thomas Moon left Bastogne with two M-16 halftracks to escort a supply train sent to the rear for supplies. While they were gone the Germans seized the roads into the town and the train was unable to return. That left Lieutenant Doby the only C Battery officer in Bastogne.

Mess Sergeant Sklenar was no less happy to be surrounded by paratroopers. He and other cooks of the 482nd had somehow…were manning a kitchen truck that was parked in Bastogne's town square. They cooked for airborne men who were streaming by until they ran out of [food] *stocks. Then they took a dozen smoked hams from a town store and sliced them up for the paratroopers.*

"I asked one guy what he was going to do without a rifle."[17] *Recalled Charles Sklenar, "He said he would take one off a dead Kraut."*[18]

Lieutenant Doby spent the day of December 21 reorganizing the battery and assigning replacements to working combat vehicles.

That evening the battery was tasked with providing thirty men to augment the 327th Airborne Infantry deployed around

---

[17] Many of the paratroopers initially arrived in Bastogne without their equipment, including their M-1 rifles.
[18] From '*Those who Hold Bastogne*' p. 45.

Bastogne in a thin defensive line. All the men not needed to man the vehicles were sent to become infantry. Eight Battery C men were cut off and trapped with their infantry comrades for eight days without relief.

Lieutenant Doby worked out a system of relief for the rest of the men in the defensive line. During one such changing of the guard the truck the men were in was strafed by a British warplane. It was thought at the time that the plane had been captured and used by the Germans.

Artillery pounded Bastogne during the day and the *Luftwaffe* bombed the town at night. The battery was down to just eight halftracks for air defense. Although it was very hard to see and track enemy planes in the darkness, C

Battery shot one of the bombers down at the edge of town on the night of the 23rd.

From December 23rd to the 26th all available men were used to shore up the 327th Airborne Infantry. On Christmas Eve night a bomb struck the aid station near the battery and all available men labored during the night evacuating the wounded and carrying water in water cans to put out the resulting fire.

Mersch, Luxembourg

When the German attack began Battery D was assigned the air defense of the command post of Combat Command R, 9th Armored Division. The battery was also assigned to defend the railhead in the town of Mersch, Luxembourg and the bridge at Colmar-Berg.

Battery D was also prepared to play a role in ground defense. To see this secondary role in action D Battery commander Captain Harry A. Ripkey visited Battery A of the battalion already providing ground defense in the town of Haller, Belgium. The captain remained there observing anti-aircraft halftracks in ground combat from 17 to 21 December.

On 22 December the 9th Armored Division moved from Mersch to the town of Etalle. Battery D was relieved of its mission and moved to Buzenol where it was used in both an anti-aircraft and a ground combat role as it supported three separate taskforces. This mission lasted until December the 26th. On that last day Battery D relieved Battery A, and was subsequently attached to Combat Command A of the 9th Armored Division near the town of Longlier.

Combat Command A, 9th Armored Division moved northeast to relieve the 101st Airborne Division, Combat Command R of the 9th Armored Division, and Combat Command B, 10th Armored Division all surrounded and besieged at Bastogne. Battery D moved in the attack as part of Combat Command A on the 27th of December.

In preparation for attack, the First Platoon of Battery D was attached to the 19th Tank Battalion. The Second Platoon was assigned a ground combat mission in conjunction with the 60th Armored Infantry Battalion.

Combat Command A was under the command of Colonel Thomas L. Harrold who divided his command into two taskforces. Taskforce Collins consisted of Second Platoon, Battery D 482nd AAA AW (SP) Battalion; Company A of the

60th Armored Infantry Battalion, Company C 19th Tank Battalion, the First Platoon of Company B 811th Tank Destroyer Battalion, and the Third Platoon of Company A 9th Engineer Battalion.

Taskforce Karsteter was composed of First Platoon, 482nd AAA AW (SP) Battalion; Company B 60th Armored Infantry Battalion, Company B 19th Tank Battalion, the Third Platoon of Company B 811th Tank Destroyer Battalion, and Company A, 9th Engineer Battalion.

The mission of these two taskforces was to seize the towns of Sibret and Villeroux, and the high ground southwest of the city of Bastogne.

At 0730 hours on 27 December Taskforce Collins moved from Longlier traveling toward the town of Sibret. At the same time Taskforce Karsteter moved out enroute to Villeroux. The outskirts of Sibret were gained at 1600 hours that afternoon. The Second Platoon, Battery D was sent to pass through Sibret to secure the towns of Jodenville and Flohomont to the west.

Two of the gun sections supported by two tanks spotted enemy activity and opened fire upon entering Sibret. The halftracks opened up with all guns firing short bursts with .50 caliber machine-guns and shelling with their 37mm cannon. The two tanks engaged the town's larger buildings with their main guns and were answered by enemy machineguns. It wasn't long before Sibret was cleared of Germans.

Two similar pairings of gun sections and tanks were organized under lieutenants Raymond Mace and Robert Yake. One group moved upon Jodenville and the other on Flohomont respectively. The tactics here were the same as those used at Sibret and both villages were soon in the hands of American soldiers. Two infantrymen were attached to each gun section and they now spread out searching the houses and rounding up prisoners. Forty enemy prisoners were taken in this well run operation that was completed by 1800 hours.

The platoon held the two towns and patrolled the connecting road that night. Patrols discovered that two German officers and a number of enemy soldiers were killed in the fighting.

Tech 5 Leon Bush was wounded during the fight in both his hands and face by shrapnel from an 88mm gun. Despite his wounds he continued his mission and guarded prisoners until relieved and evacuated for treatment.

While the above action was underway, the First Platoon of Battery D was engaged with the enemy at Villeroux as part of Taskforce Karsteter.

First Platoon was providing air defense for the bridge crossing that led into the town and then took up flank guard as the taskforce moved through Villeroux. The First Gun Section destroyed a German occupied house and then captured the enemy soldiers as they tried to flee.

The platoon came under enemy artillery fire that afternoon.

While searching house to house in Flohomont Lieutenant Davis liberated an Allied pilot shot down and captured by the Germans as he was flying supplies in to Bastogne. The pilot said the Germans were terrified by the M-16 'secret weapons' that chewed them up the day before.

The twin towns were subjected to 88mm gun shelling during the day resulting in Tech 5's Joyce and Oberster being wounded. Joyce was treated and returned to duty but Oberster's wounds were more serious and he was medically evacuated.

A total of forty six more prisoners were taken by Battery D on the day of December 28th.

In the afternoon December 29 the Second Platoon was ordered to leave the twin towns to attack and secure the high ground southwest of the town of Chenogne. The platoon, with four tanks attached, moved out at 1300 hours and quickly came under mortar and sniper attack. Infantry was brought forward to clear the enemy positions. The attack faltered after about 800 yards as German artillery knocked out one of the halftracks.

Lieutenant Yake assisted as forward observer for the 3rd Field Artillery and directed fire upon enemy anti-tank guns. Although Chenogne was taken it could not be held in the face of the enemy counterattack and the small taskforce withdrew and dug-in south of the town.

The platoon's mission was changed to defending the town of Villeroux and during subsequent fighting Corporal Dobson and Tech 5 Capiglioni were wounded.

At dawn on the 30th of December the Germans launched a counterattack in an effort to cut the road leading to Bastogne. The Second Platoon withdrew and set up a defensive line stretching from the town of Jodenville to the village of Sibret.

The enemy assault was stopped and the Second Platoon set up a defense on the north edge of Sibret to cover elements of the 60th Armored Infantry Battalion. The infantrymen were dug-in near a large wooded area north of the town and the platoon provided an over-watch for them.

During the fighting near Jodenville a Sherman tank was knocked out by enemy fire and an escaping wounded crewman fell under fire. Staff Sergeant Nicholson of Second Platoon halted friendly fire and Private First Class Velmer Huntsman low crawled to the wounded man and dragged him to safety.

At break of day on 31 December the Second Platoon held the north side of Sibret and the First Platoon still held the road to Villeroux. Intermittent 88mm gunfire and mortar barrages were all the enemy could muster that day.

At 1100 hours on New Year's Day of 1945, orders were received to attack in conjunction with elements of Company A, 60th Armored Infantry Battalion. The mission was to finally clear the Germans oc-cupying the woods in front of the armored infantrymen.

Fifteen minutes after receiving the orders M-15s and M-16s of the platoon opened fire on the woods as the infantry advanced. The heavy machinegun fire and 37mm guns lit the forest up in red tracers and hot lead explosions. The heavy fire ripped up enemy positions and it was not long before the Germans began to give ground. In short order nearly a hundred of the enemy were killed or wounded and more than a hundred quickly surrendered.

Combat in the 482nd AAA AW Battalion's area of operations came to a halt. The next two days saw Battery D hold and maintain its defensive positions until ordered on 3 January to move to an assembly area near the town of Witry with the rest of Combat Command A.

The next day, the 4th of January 1945, the battery was relieved from its attachment to Combat Command A, 9th Armored Division and reverted to battalion control. The battery moved to the vicinity of Balaives, France for refitting and reconstituting with replacements as needed.

On January 8, Battery B was relieved from its attachment to Combat Command B, 9th Armored Division and sent for refitting to the area of Sapogne,

France. But the very next day its Second Platoon was assigned to defend the bridges near Les Bulles, Belgium.

Also on the 8th of January the 482nd AAA AW (SP) Battalion was relieved from the 9th Armored Division and attached to VIII Corps under the 113th AAA Group and moved to Jamoigne, France.

On the 9th of January Battery A was assigned AA protection of a supply dump in the vicinity of Marbehan, Belgium.

Other minor shifts of positions and defense missions continued for the next week until January 12 when the 482nd AAA Battalion was relieved from VIII Corps and attached to the XX Corps under the 112th AAA Group.

Although this was a 'busy' time as the batteries of the battalion were shifted from place to place to provide air defense, it was also a welcome break and allowed the men to get some badly needed rest. The vehicles were serviced and repaired and the entire battalion brought back to its full fighting potential.

Notable personnel changes were 1st Lieutenant David Doby was placed in command of Battery C, and Captain Milton Pullen was assigned as battery commander of Headquarters Battery.

Also during this time the separate batteries and the battalion received no less than six well-earned commendations from higher headquarters for its service during the German Ardennes Offensive of December 1944 and January 1945.

Germany

February continued in a similar vein as January but the battalion found itself once again attached to the 9th Armored Division. This assignment seemed to always be a harbinger of hot operations to come. One notable change was that First Lieutenant William Higgins assumed command of Battery A.

The other notable event was that on February 28 the 482nd AAA AW (SP) Battalion moved into Germany. They would now take the fight to the hated enemy's homeland.

All batteries of the battalion were now attached to separate commands of the 9th Armored Division.

Battery A went to Combat Command A, with its Second Platoon attached to its headquarters, and the First Platoon further attached to the 3rd Field Artillery Battalion.

Battery B was attached to Combat Command B with the First Platoon further attached to the 656th Tank Destroyer Battalion. The Second Platoon was attached to the 16th Field Artillery Battalion.

Battery C was assigned to provide AA protection for the division supply trains.

Battery D was attached to Combat Command R and the First Platoon further attached to the 73rd Field Artillery Battalion.

It was in this configuration that the 482nd AAA Battalion entered and proceeded forward toward the heart of Germany.

On March 2 B Battery engaged two ME109s near Muddersheim with both

37mm cannon and .50 caliber machine-guns. The two aircraft sheared off with one of them trailing smoke. Neither plane caused any damage or casualties.

Except for light shelling from time to time the 9th Armored Division's advance was hardly contested. The separate battery's missions underwent minor changes as the days passed.

However, on March the 7th the Lundendorf Bridge at Remagen fell mostly intact to the 9th Armored Division. This was a major mistake by the Germans, and although the damaged bridge would later collapse, it would be too late as the Americans rushed across and gained a foothold on the other side of the Rhine River.

To Save a Bridge

The bridge—large and ugly—and the river loomed ahead. Floor planking was laid over the tracks on the bridge to convert the span from railroad to vehicular traffic. The river ran swift and deep between the towering cliffs. No one knew the strength of the enemy on the other side of the river or just when the explosions would be touched off.

Lieutenant Timmerman gathered his forces near the bridge and gave them their instructions. It was 1550. In the face of murderous fire the 27th Armored Infantry had just 10 minutes to cross the river if the Germans were running on schedule.

The *New York Sun* would later say:

"The Germans misjudged by a fateful ten minutes the speed at which the 9th

Armored Division was moving... To all who utilized that ten minutes so advantageously goes the deepest gratitude this country can bestow." [19]

The bridge was captured at 1600 hours and there was little doubt the Germans would try to correct their error by sending bombers against the long span of the bridge.

Remagen Bridge as seen from an aircraft

[19] Taken from a Stars and Stripes article.

The closest anti-aircraft artillery unit was the 482nd AAA AW (SP) Battalion. By 1900 hours D Battery was assigned the vital mission of air defense and was soon enroute. Batteries A and B were soon assigned the same mission. Night had fallen but the three units made their way as quickly as possible to the bridge.

By 0200 hours A Battery under Lieutenant William F. Higgins, was in the German town of Ringen and in position to defend the bridge at Remagen. Higgins was commissioned in the Coast Artillery in 1939 but by 1942 he was an anti-aircraft artillery officer.

Suddenly a heavy concentration of mortar fire began to fall in and around the town. During the barrage Private Thomas Torchie was struck and killed by shrapnel.

Despite the continuing barrage, First Lieutenant James French, the assistant platoon leader of A Battery's Second Platoon, ran to Private Torchie and evacuated him to the aid station.

Lieutenant French was then ordered to take his platoon across the bridge and join the growing bridgehead. Lieutenant

Battery A, 482nd was the first AAA unit to defend at Ludendorff Bridge.

French and his Second Platoon of A Battery would be the first 482nd unit to arrive on the east bank of the Rhine River.

By 0645 hours Lieutenant Keogan had the Second Platoon of A Battery set up for the air defense of the bridge. Lieutenant Winchester with his Second Platoon of B Battery was also ready and likewise Lieutenant James Davis' D Battery was up. Other 9th Armored units were in combat in the nearby town of Erpel.

Heavy enemy artillery of 155mm and larger began falling upon the town of Remagen and the bridge at 0811 hours.

In the meantime all other platoons of the 482nd except Battery C, arrived and took up defensive positions around the bridge, ready for the coming air strike.

Two more platoons, the First Platoon of A Battery under Lieutenant Gilbert Barnes, and the Second Platoon of D Battery under Lieutenant Robert Yake crossed the bridge to the east bank. All units were now in position.

The air was thick with anticipation of the German air attack sure to fall soon.

Ludendorff Bridge at Remagen

However it was surprisingly late in the afternoon before the Luftwaffe arrived.

The Germans increased their artillery barrage on and around the bridge and in the river and then the air attack began.

Major Ben Cothran, the S-3 officer for Combat Command B, 9th Armored stated

That afternoon the first planes came in for the bridge. But Pappy (Captain Denton), *who had done wonders in the Bulge, had got his antiaircraft battery together somehow, brought it through all the snarled traffic and was there. Because of high hills . . . the planes had to make a straight run up the river for the bridge. Four tried it that afternoon. Pappy's lads got all four.*[20]

[20] From: "The Hammer of Hell" Cothran attributed the AA fire to Battery A, however Battery D was also engaged.

Sortie after sortie of enemy aircraft dive bombed the targeted bridge and strafed ground units. Each plane came in at an altitude of two thousand feet, dropped down between five hundred and two hundred feet and then dropped their bombs. HE-111s, ME-109s, and JU-87s each took their turn. One ME109 was destroyed as well as three JU-88s in the first attack by the 482nd AAA AW Battalion. An HE-111 sped away as a piece of its wing was shot off. All bombs missed their targets and damage by either bombs or artillery barrages was negligible. There was no vehicle damage or casualties among the ranks of the battalion.

At the end of the day all batteries except C Battery were stationed at each end of the Ludendorff for the best possible protection of both the bridge and the other 9th Armored Division units guarding it.

The battalion fired more than 30,000 rounds of .50 caliber ammunition as well as 253 37mm high explosive shells.

Battery C remained protecting the division's supply train of men and materiel.

American forces continued to cross over the damaged railroad bridge as sporadic enemy artillery barrages crashed down around them.

As was expected, enemy air activity continued but was inef-fective as the guns of the 482nd AAA AW continued to drive them away.

In the afternoon the batteries got a good look at a jet propelled plane identified as a German Arado AR-234 flew by out of gun range.[21]

[21] The world's first operational jet powered bomber. It was built by the Arado company near the close of the war.

A German Arado AR-234

There were so few of these aircraft that they were used primarily for reconnaissance. This one did not bomb or strafe.

The total rounds expended in defense of the bridge on March 9 were more than 23,300 rounds of .50 caliber and 403 37mm shells. One M16 of the Second Platoon, B Battery was damaged by artillery fire killing Lieutenant Frank W. Winchester who was nearby directing anti-aircraft fire during the barrage.

1st Lieutenant Frank W. Winchester

This same artillery round also killed Sergeant Lamberson; the squad's section leader, while he was delivering ammunition to the gun.

Tech 5 Boyce of the same halftrack was severely wounded and later died of his wounds at the aid station.

That day was also hard on Battery A as Corporal Cancelliere and Tech 5 Young were wounded in action while Tech 5 Lafernier and Private Patrone were both killed. All these men were steadfastly in the turrets of their guns when they were hit. Later in the day an M-16 of the battery was knocked out.

That night the combat engineers began patrolling the river in boats near the Remagen Bridge. These engineers were dropping prepared charges at intervals in the river to deter enemy divers suspected of trying to place underwater explosives on the bridge. The following day they placed nets to keep both floating objects and or swimmers away from the base of the bridge.

In the early hours of March the 10th, B Battery lost an M-16 to artillery fire, but no casualties resulted. During the day no less than ten enemy aircraft attacked the bridge with five claimed by battalion fire.

The planes came in at various altitudes between five hundred and thirty-five hundred feet using evasion maneuvers. The battalion's AA fire kept the bombers from approaching in anything resembling a bombing run. They did not attack in formation but by individual plane. They dropped bombs and strafed with little effect. One M-16 halftrack and two M-15s were damaged during the day. Tech 5 Frank Roberts of B Battery was wounded in action.

On March the 11th the three batteries of the 482nd continued its AA defense of the Ludendorff Bridge.

A treadway bridge across the Rhine.

A half mile north of the Ludendorff Bridge the combat engineers completed a heavy treadway bridge across the Rhine to increase the flow of American forces. The bridgehead meantime was growing in depth and width and the U.S.

Army had a firmer control over the mighty river.

An FW-190 strafed this new bridge killing one officer and a war correspondent. However, the plane was destroyed and the pilot killed by AA fire. Captain Denny of B Battery reported a V-1 and D Battery reported three V-2 flyovers headed west.[22]

That evening Battery A destroyed an enemy plane and the pilot was captured. Under questioning he willingly gave up the location of his airbase.

Tech 5 Aubrey Fullen of D Battery was wounded in action and one M-16 belonging to B Battery was damaged.

The bridgehead on the east bank pushed further east, driving the enemy artillery out of range of the bridge. As a result the enemy fire was even more

---

[22] A V-1 was a winged cruise missile while a V-2 was a finned rocket that flew much faster than the V-1.

inaccurate and less frequent. The Americans were winning the bridgehead. The enemy air attacks were ineffective; as they were winning the battle for the skies too.

At 1500 hours that afternoon B Battery was relieved and sent to support the 667th Artillery. Battery D, 639th Automatic Weapons Battalion took B Battery's position near the bridge.

The enemy seemed to change from primarily flying FW-190s to ME-109s possibly because most of the former model of plane was destroyed. The batteries continued to expend ammunition at a fairly high rate with more than 34,000 rounds of machinegun ammo and 102 37mm shells.

That same day D Battery was replaced by Battery C of the 639th. As D Battery pulled out in column it was dive bombed by an FW-190. Four men, Tech 4 Scelina and Tech 5 Sumpter as well as privates first class Eddy and McMahn were all wounded in action and evacuated.

At the end of the day of March 12, the 482nd AAA AW (SP) Battalion was in the following configurations.

A Battery continued defending the bridge, B Battery was under battalion control for refitting at Heppingen, C Battery was in defense of the 9th Armored Division trains, and D Battery was in AA defense of the 16th Field Artillery at Heppingen.

On March 15 the 482nd was relieved from all bridge defense by the 639th AW Battalion.

The excitement of defending both the Ludendorff and treadway bridges was over as the bridgehead was now seven miles long and fourteen miles wide. The 482nd AAA AW (SP) Battalion resumed its primary mission of air defense with the 9th Armored Division.

A recap of the battalion's actions during the period of bridge defense shows the following statistics.

16 enemy planes were destroyed and three heavily damaged. 6 other planes were destroyed in conjunction with fire from other units. During the engagement a total of 303,337 rounds of .50 caliber machinegun ammunition were fired as well as 4,664 of the 37mm gun's shells.

The battalion lost 4 M-15s and 7 M-16 halftracks. Four men were killed in action, two more died of their wounds, and twelve were wounded in action. Of course mere statistics is not an accurate measure of each soldier's pain and suffering both physically and mentally.

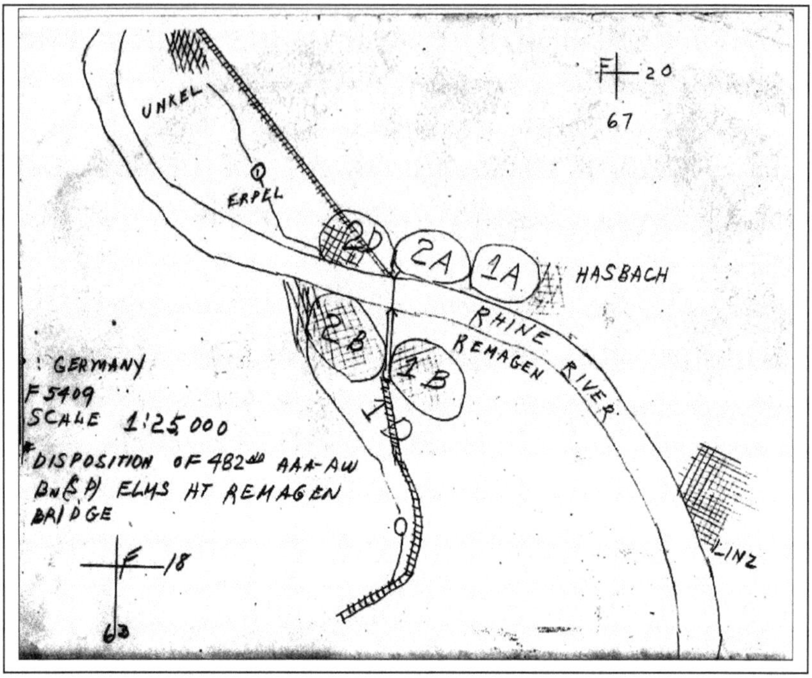

A hand drawn map showing the dispositions of the 482nd while assigned to the Remagen Bridge.

The enemy launched 53 air sorties involving 79 planes of which 22 were destroyed by the 482nd and 3 were damaged.

An interesting note on the chaos of battle is what happened during the first few days on the streets of Remagen. The 9th Armored Division's two armored infantry battalions crossed the bridge on foot; they left vehicles behind such as jeeps and halftracks with machineguns, sitting in the city's streets.

Later passersby would jump on these vehicles when any plane appeared and begin firing almost indiscriminately. Red tracers and lead soon filled the skies but also filled the streets of Remagen with dangerously ricocheting bullets.

Eventually small arms were used including pistols, M-1s, and Tommy guns on any aircraft. Friendly planes often came under fire from troops not trained in anti-aircraft fire or in the identification of planes. The friendly pilots were blaming the AA soldiers when they hadn't in fact fired a shot at them.

Quickly an order came down that no one who was not an AA soldier was to fire at planes unless their unit came under direct attack by an aircraft.

For the next week or so the 482nd was given a welcome respite from combat of any sort. Weapons were cleaned and maintenance, training was conducted not only for the new men but the veterans as well. All the men got refreshing showers with hot water, clean uniforms, and some badly needed rest.

A German ME-262 jet powered fighter-bomber.

By the 21st of March it was obvious that something was in the wind. A Battery was attached to 9th Armored Division Artillery and B Battery to the division's Combat Command B.

On the 23rd A Battery received some inaccurate and ineffective enemy artillery fire. An indication the war was fast moving toward a conclusion was when the battalion received an alert that American C-47s with Russian markings would be operating in our area.[23]

The battalion was now shifted under V Corps and the 115th AA Group. Two days later the 482nd mission was changed from air defense to offensive ground operations. The battalion rolled forward with the 9th Armored Division as that unit made a swift and deep penetration into Germany. The 482nd became an integral part of the encirclement of the industrial 'Ruhr Pocket' resulting in the isolation and capture of thousands of enemy troops and their equipment. The war had just been shortened and future casualties reduced.

On the 27th of March B Battery, attached to Combat Command R, jumped off to the south of Limburg,

[23] The C-47 was the U.S. standard cargo plane.

Germany to establish contact with the 3rd U.S. Army.

The First Platoon advanced with Taskforce Schantz and Second Platoon with Taskforce Devers. At 0930 hours D Battery was relieved from attachment to Combat Command B and sped forward to protect the bridges and underpasses of the autobahn the division was traversing.

B Battery was in position and ready to fire by 1130 hours. Other elements of the 9th Armored were meeting slight opposition although no enemy force seemed willing to oppose the 482nd with their heavy weapons. At the end of the day the battalion was located in the towns of Elz and Altendiez.

On the following day B and D Battery were used to flush out foxholes and small wooded areas, expending a little over 3,000 rounds of .50 caliber ammunition.

At 1830 hours D Battery was relieved from bridge protection and reverted to Combat Command B.

The next morning on the 29th of March the battalion continued to advance to the north with the 9th Armored. Thousands of prisoners were taken but no opposition was encountered, it appeared that Germany, or at least the western half of Germany had accepted defeat.

At 1800 hours an ME-109 attacked and strafed Combat Command A's armored column, A Battery engaged the enemy plane who promptly broke off the attack. A second aircraft attacked the 19th Tank Battalion but also broke contact when D Battery opened fire. Units of the battalion reported the plane was smoking and losing altitude the last it was seen.

Another ME-109 attacked the observation plane of the 16th Field Artillery Battalion near Momberg. As soon as the batteries engaged the plane peeled away, its attack also thwarted.

D Battery engaged enemy troops at the edge of a wood killing one, wounding four and capturing twenty-seven prisoners. All this was accomplished with the expenditure of a mere 125 rounds of .50 caliber ammo.

On the 31st of March D Battery joined with two tanks of Company A, 17th Tank Battalion to attack the town of Bergheim. There was a brief fight with some SS troops and a couple of dug-in gun positions and the town was secured.

The battery then supported the 52nd Armored Infantry Battalion in capturing the three towns of Bad Willingen, Ossendorf, and Wethen. This victory was accomplished despite the enemy engaging them with panzerfausts, dug-in guns, and infantry.

The 9th Armored Division now linked up with units of the 9th U.S. Army and enclosed two complete German armies of over 327,000 men. The industrial might of the Ruhr valley was now in the hands of the Allies.

In this important action the 482nd had not only performed its primary mission of air defense in a superior manner, but also demonstrated it was a potent ground assault unit. Commendations from higher headquarters flowed into the battalion's message center.

Another view of Battery A at the Remagen Bridge

482nd Tracks on the Rhine near the Remagen Bridge

482nd men scan the sky for aircraft

Other AA troops at Remagen

This view of the bridge from the town looks deceptively peaceful

Above is a drawing of an M-16 based on actions of the 482nd AAA AW (SP) Battalion at Remagen Bridge. From a graphic featured in the magazine ADA Yearbook, it is believed to be in the public domain.

April 1945 the Final Push

"This victory of Allied arms is a fitting prelude to the final battle to crush the ragged remnants of Hitler's armies of the west, now loitering on the threshold of defeat."

General Dwight D. Eisenhower

The 9th Armored now assumed a spearheading role, leading the way for First Army's drive to the east. This final drive through the heart of Central Germany began on April 10. Division tanks smashed so deeply into the enemy's rear that German units became hopelessly confused. Communications were slashed, and vital supply lines were severed.

In these April operations, the 9th Armored Division's combat commands advanced almost three hundred miles from the town of Warburg to the Mulde River while completing the encirclement of the city of Leipzig. Attacking abreast, the three combat commands captured hundreds of towns and villages, and took thousands of prisoners while destroying enemy tanks, guns, and vehicles. It was a rout.

Throughout this unprecedented advance the 482nd AAA AW (SP) Battalion continued to play their role of support both in AA defense and ground operations.

The 9th Armored Division's drive cut Germany in two. Instead of continuing the drive east to meet the Russians, the division was pulled out of the line for a well-deserved rest.

There was one remaining task when the enemy threatened a continued resistance in Czechoslovakia. The 9th Armored Division with the 482nd providing security, pushed on with the First Infantry Division. By the time the link-up was made with the Red Army in Karlsbad, the Germans were finished.

Occupation Duty

The war in the European Theater ended on May 8, 1945 with the formal surrender of all German forces. At the time the 482nd was still attached to Combat Command B, 9th Armored Division which itself was attached to the 1st U.S. Infantry Division[24].

Battery A was astraddle the border of Germany and Czechoslovakia with a platoon in the village of St. Sedlo and in the town of Weiden, Germany. The rest of the batteries were still inside Germany.

In the following days the combat command was assigned an area of occupation and the 482nd remained attached to them and were located at the following towns in Germany: Battery A

[24] The author served in Battery A, 1st Battalion, 7th Artillery, 1st Infantry Division in Vietnam.

at Weissenstadt; Battery B at Brand, Battery C at Arzberg, and Battery D at Marktleuthen.

Each battery set up road blocks and took over prisoner of war cages already constructed by the 16th Armored Field Artillery Battalion.

In the following weeks and months the battalion was shifted a number of times to different areas of Germany.

No doubt the men of the battalion heard the usual rumors that the unit would remain on active duty and eventually be sent to the Pacific Theater. As the reader is aware however, the war ended in August of 1945 with the surrender of Japan.

The U.S. Army in Germany was swiftly cut and units sent back to the U.S. and processed out of the Army.

However some units like the 482nd remained on occupation duty.

The 482nd AAA AW (SP) Battalion was finally returned home and mustered out of active service on 30 April 1946 at Camp Kilmer, New Jersey. They had missed the ticker tape parade in New York City. The unit would continue to exist in the Army Reserve until 1959 when it was finally placed on the inactive unit list.

In January 1946 a victory parade for WWII units was held four months before the 482nd AAA came home.

The 482nd AAA AW (SP) Battalion had served exceptionally well, with half of its batteries earning individual Unit Citations. The citation for Battery A was issued as General Order 44, from General George C. Patton's Headquarters, Third U.S. Army on 22 February 1945.

Battery A, 482nd Anti-Aircraft Automatic Weapons Battalion (SP) is cited for its action during the period 17 December – 26 December 1944. During this period Battery A was assigned the mission of furnishing ground support to infantry, artillery, and tank destroyer units defending positions against the strong enemy thrust into Belgium and Luxembourg. Facing overwhelming odds with outstanding heroism the battery fought stubbornly over difficult terrain, inflicting heavy losses upon the enemy. At one time this unit held a sector alone, without the immediate support of other arms! The indomitable fighting spirit; the courage and devotion to duty of the officers and men of this battery are in keeping with the highest traditions of the military service.

The citation for Battery C was issued as General Order 31, from General Patton's Headquarters, Third U.S. Army on 7 February 1945.

Battery C, 482nd Anti-Aircraft Automatic Weapons Battalion (SP) is cited for its action during the period 18 December – 27 December 1944 for extraordinary heroism and gallantry in the defense the key communications center of Bastogne, Belgium. Essential to a large scale exploitation of his breakthrough into Belgium and northern Luxembourg, the enemy attempted to seize Bastogne by attacking constantly and savagely with the best of his armor and infantry. Without benefit of prepared defenses, facing almost overwhelming odds and with very limited and fast-dwindling supplies, [this unit] maintained a high combat morale and an impenetrable defense, despite extremely heavy bombing, intense artillery fire, and constant attacks from infantry and armor on all sides of their completely cut-off and encircled position. This masterful and grimly determined defense denied the enemy even momentary success in an operation for which he paid dearly in men, material and eventually morale. The outstanding courage and resourcefulness and undaunted determination of this gallant force is in keeping with the highest traditions of the service.

Presidential Unit Citations

These Third Army unit citations with minor changes to the narrative would later be approved by the War Department as Presidential Unit Citations. This action authorized each member of the unit to wear the distinctive blue ribbon with gold metal frame of the award.

Battery B received a Presidential Unit Citation while serving as a part of Combat Command B, 9th Armored Division. That citation reads as follows.

Combat Command B, 9th Armored Division is cited for extraordinary heroism in combat in the vicinity of St. Vith, Belgium, from December 17th to December 23, 1944. Combat Command B, 9th Armored Division, was subjected to repeated tank and infantry attacks, which grew in intensity as German forces attempted to destroy the stubborn defenses that were denying them the use of the key communication center at St. Vith. By the second day, the flanks were constantly threatened by enemy forces that had bypassed the St. Vith area and were pushed far to the rear in an effort to encircle the command east of the Salm River. The attacking forces were repeatedly thrown back by the gallant troops who rose from their foxholes and fought in fierce hand to hand combat to stop the penetrations and to inflict heavy losses on the numerically superior foe. As the command continued to deny the important St. Vith highway and railroad center to the Germans, the entire offensive lost its initial impetus, and their supply columns became immobilized. By 21 December, the German timetable was so disrupted that the enemy was forced to divert a corps to the capture of St. Vith. Under extreme pressure from over-whelming forces, Combat Command B. which for seven days had held the St. Vith area so gallantly was ordered to withdraw west of the Salm River. By their epic stand without prepared defenses and despite heavy casualties, Combat Command B, 9th Armored Division, inflicted crippling losses and imposed great delay upon the enemy by a masterful and grimly determined defense in keeping with the highest traditions of the Army of the United States.

The entire battalion was authorized to display a campaign streamer on the unit's regimental flag for the PUC embroidered *Remagen Bridgehead.*

Battery A was authorized to display a streamer on their battery guidon embroidered with the word *Ardennes.* Battery C was authorized to display a streamer on their guidon for the PUC embroidered *Bastogne.* Battery B was authorized a streamer for the Belgian Croix-de-guerre with Palm for action in *Bastogne* and the *Ardennes* as follows:

Battery B, 482nd AAA Automatic Weapons Battalion (SP) [is] cited in the Orders of the Day of the Belgian Army, by Decree Number 7253, 13 July 1950, by Charles, Prince of Belgium, Regent of the Kingdom.

It should be noted here that these three batteries, fighting independent of one another and in different locations during the Battle of the Bulge, rendered service so deserving as to receive these high honors as individual units.

The 482nd is credited with serving during three campaigns in Europe including *Rhineland, Ardennes-Alsace,* and *Central Europe.*

The men of the battalion were authorized to wear and display three bronze campaign stars on their European - Middle Eastern campaign medal.

A letter of commendation also came from General Omar Bradley, commander of the 12th Army Group to the commander of the 9th Armored Division and reads:

Will you convey to General Leonard and the troops of the 9th Armored Division my greatest admiration for the skill and daring with which they have crossed the Rhine to seize and establish the first bridgehead for the Allied armies. This bold advance characterized by able willingness to chance great risk for great rewards speeds the day when our full forces can come to grips with the enemy.

Enemy Aircraft

A JU88 night fighter-bomber

An ME-262 jet over Belgium

A formation of ME-109 fighters

An FW-190 fighter

One Soldier's Story, Corporal Charles W. Adamek

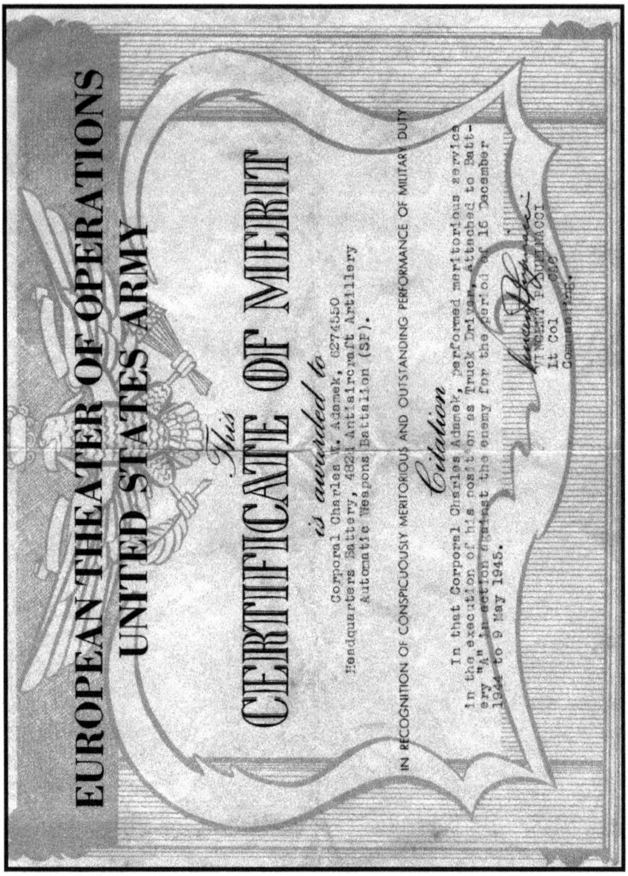

EUROPEAN THEATER OF OPERATIONS
UNITED STATES ARMY

This
CERTIFICATE OF MERIT
is awarded to

Corporal Charles W. Adamek, 6274550
Headquarters Battery, 482d Antiaircraft Artillery
Automatic Weapons Battalion (SP).

IN RECOGNITION OF CONSPICUOUSLY MERITORIOUS AND OUTSTANDING PERFORMANCE OF MILITARY DUTY

Citation

In that Corporal Charles Adamek, performed meritorious service in the execution of his position as Truck Driver, attached to Battery "A" in action against the enemy for the period of 16 December 1944 to 9 May 1945.

VINCENT P. [illegible]
Lt Col CAC
Commanding.

Meritorious Service Certificate for Charles W. Adamek.

The record of one 482nd AAA AW Battalion soldier was researched at the request of a relative who forwarded the information to the author. It is included here as a look at a *typical* member of the unit, all of them from the Greatest Generation.

Charles W. Adamek enlisted into the U.S. Army in 1941 and was initially assigned to the 76th Field Artillery. He was later sent to the 79th Coastal Artillery. He joined Headquarters Battery of the 482nd Coastal Artillery Battalion in May of 1943 when he was reassigned from the 79th CA. The 79th CA may have been on the same military base as the 482nd stationed at Camp Hulen, Texas. Charles was court martialed shortly before this, probably for going AWOL.[25] It looks like he got

[25] Absent Without Leave. This was one of the most common infractions of military discipline.

into trouble a few times. Charles was fined while he was still in the 79th Coastal Artillery Battalion. Charles was one of the first men assigned to the new battalion.

He was assigned to the 482nd Coastal Artillery in May of 1943 and later that month the 482nd Coastal Artillery Battalion was redesignated as the 482nd AAA AW (SP) Battalion. His duty code at the time shows he was training to be a radio operator.

Overcoming his past record Charles was promoted to specialist or Tech 5 on 22 September. He was on furlough in late December and must have come back late because he was charged with being AWOL on 6 January, 1944. He was subsequently reduced back down to the rank of private.

In August he sailed overseas with his new unit from Camp Kilmer, New Jersey. In September he landed on Utah Beach in France with the 482nd AAA AW (SP) Battalion.

Charles was promoted to Corporal on 16 November 1944 and his duty position at that time was as a halftrack driver.

In January of 1945 he is shown assigned to A Battery as a truck driver. On 2 April he was sent to the hospital for a non-combat related issue. The report shows his injury as *NBC* which meant non-battle casualty. He returned to the unit on 17 May. These notes were kept for benefits purposes so it was important to detail whether the injury was sustained in combat. This was the only time he was in the hospital according to the official records.

In July he was reduced to the rank of private again for some infraction, no reason for the demotion was given but was most likely absent without leave again. It was hard to ignore all the temptations Europe had to offer. He was promoted back to Private First Class in August.

Charles had a 7 day furlough in October and then on the 17th of October was assigned out of the 482nd for transport to the United States.[26] There would seem to be little doubt he was happy to end his military service and resume his life as a civilian. His friends in the 482nd would not follow for another six months.

However, Charles must have missed his time in the Army because he decided to reenlist and did so in 1947. This time

---

[26] Research on Charles Adamek paid for and provided by his nephew and contributor to this unit history, Mr. Jim Lefemine.

he made a career of the military and served until his retirement in 1962 at the rank of Sergeant First Class.

482nd Officer Roster at War's End

Lieut. Col. Vincent P. Lupinacci, Battalion Commander

Major Maddox, Executive Officer

Captain Pulliam Battalion S-4 Officer

First Lieutenant Ness, Battalion Adjutant

Lieutenant Laplue Battalion Motor Officer

Captain Herman A Battery Commander
First Lieutenant Keogan Executive Officer

Lieutenant Oakes 2nd Assistant Platoon Leader

Captain Denton B Battery Commander
First Lieutenant Woodrow Willard Executive Officer
First Lieutenant Frank Winchester 1st Platoon Leader
Lieutenant David Elias Assistant Platoon Leader
First Lieutenant Lou Jared 2nd Platoon Leader
Lieutenant Warren Haick Assistant Platoon Leader

Captain Denny Lovoi C Battery Commander
First Sergeant Martin
First Lieutenant David Doby Executive Officer

First Lieutenant Carlton Smith 1st Platoon Leader
Lieutenant Gomer Cullifer Assistant Platoon Leader
First Lieutenant Jim Crabb 2nd Platoon Leader
Lieutenant Jacob Gerber Assistant Platoon Leader
Lieutenant Thomas Moon Assistant Platoon Leader

Captain Harry Ripkey D Battery Commander
First Lieutenant Jim Davis Executive Officer
Lieutenant Bob Yake 1st Platoon Leader
Lieutenant Tom Ferguson 2nd Platoon Leader

Author's Afterword

Red Tracers is my fourteenth unit history either originally written by me or edited by me from a WWII unit history 'handbook' issued to soldiers at the end of the war. These unit histories including this book are too important to let fade away like old soldiers. In writing these short narratives it is my intent to keep the deeds of these men available to the public and the veterans themselves and even more importantly, available to their descendants.

If I have accomplished this task for the 482nd I will be satisfied that I have made a contribution both as a writer and a historian. I hope you found the effort of reading this book to be worthwhile.

The 482nd Anti-Aircraft Artillery (Automatic Weapons) Battalion (Self Propelled) Unofficial Unit Insignia.

About the Author

Walt Cross is a retired U.S. Army master sergeant. He is a *Pheon* veteran of the Vietnam War and served with Battery A, 1st Battalion, 7th U.S. Artillery[27], 1st Infantry Division from 1968 to

[27] When the First Infantry Division was alerted for deployment to Vietnam, the then 7th Field Artillery was redesignated the 7th Artillery and kept that designation throughout its five years of combat in Vietnam. Its unit insignia was also changed and because of the three pheons (arrowheads) on the shield, the men were known as *Pheons.*

1970. While with the division he was awarded the Bronze Star Medal twice, once for heroism in combat and once for meritorious service.

Walt holds university degrees in both history and science. Additionally, he is a graduate of the U.S. Army Sergeants Major Academy (located at Fort Bliss). He is also the Executive Director emeritus of the Oklahoma ESGR (Employer Support of the Guard and Reserve) a Department of Defense soldier support program.

He is a proud father and grandfather and he resides in Stillwater, Oklahoma with his wife Carol.

CPSIA information can be obtained
at www.ICGtesting.com
Printed in the USA
LVHW081147311219
642042LV00023B/711/P